Hamshack Raspberry Pi

Learn How To Use Raspberry Pi For Amateur Radio Activities And 3 DIY Projects

Introduction

I want to thank you and congratulate you for purchasing the book, *"Hamshack Raspberry Pi: Learn How To Use Raspberry Pi For Amateur Radio Activities And 3 DIY Projects"*.

This book has actionable information on Hamshack Raspberry Pi that will help you to learn how to use Raspberry Pi for amateur radio activities and much, much more.

Since you are reading this book, I'm sure you are enthusiastic about radio activities, if not about the revolutionary Raspberry Pi. I'm also sure there's so much you've heard about radio stations like tracking satellites, communicating in Morse code or perhaps playing a game over the air as well- and you want to try them out.

That's very possible, and you don't have to spend a lot of money to learn all that- and become a professional radio operator- because a cheap raspberry Pi computer and a couple of other cheap tools are all you require to begin your amateur radio journey.

Maybe you haven't been adequately introduced to the small computer known as Raspberry Pi; that is not a reason to skip reading this book because you will know everything –from the basics- about Pi, before we get to the actual playing with the Hamshack Raspberry Pi. Among other things, you will learn how to install, configure and use the device to enjoy some of the coolest things in tech today. For about $40, you will be able to enhance your knowledge of how to operate radio as an amateur; you will learn how to install different operating aids like time keeping, logging, Morse code practicing and satellite tracking. You will also learn about designing antennas, essential Ham programs like twclock and GNU radio companion, radio configuration tools and even how to set up your own ground station with simple steps!

Best of all, you'll be able to complete the projects discussed in the book by yourself without any problems because they are so damn easy and straightforward. Shall we begin?

Thanks again for downloading this book. I hope you enjoy it!

content of this book has been derived from various sources. Please consult a licensed professional before attempting any techniques outlined in this book.

By reading this document, the reader agrees that under no circumstances are is the author responsible for any losses, direct or indirect, which are incurred as a result of the use of information contained within this document, including, but not limited to, —errors, omissions, or inaccuracies.

Table of Contents

Let's start from the beginning i.e. understanding the Raspberry Pi before we get to the point of discussing various other issues surrounding the Raspberry Pi.

A Comprehensive Background of the Raspberry Pi

What Is It?

In simplest terms, the Raspberry Pi refers to a series of small computers (in the category of single board computers) that were developed in the UK by the Raspberry Foundation to help teach basic computer science in various schools both in the developed and developing countries. Out of the box, the new device, which you can purchase from Raspberrypi.org, comes without the peripheral devices (mouse, case and keyboard).

Before we discuss the specifics of how to set up and use the Raspberry Pi as a pro even as a complete beginner, we will start by going through the journey through time i.e. how the Raspberry Pi came into being. This short history will help you to understand Raspberry Pi well before we get to set it up.

A Short History of Raspberry Pi, And the Setup

Before raspberry Pi was invented, personal computers had become an expensive household appliance. Many parents had grown reluctant of letting their kids use the family computers due to the high cost of the machines and fragility. Many kids, as a result, were not well-versed with computers. In 2006, Dr. Eben Upton together with his associates from the University of Cambridge realized that there was a steep decline in numbers and skills of the students enrolling for computer science courses. They decided to develop an inexpensive computer that would enable young people familiarize themselves adequately with computer concepts.

In 2011, the Raspberry Pi Model B was created and it sold more than two million units within a period of two years. Since then, there has been an ongoing improvement in different models of Raspberry Pi.

The machine is not only a computer but a microcontroller as well with pins that can sense externally and actually control devices. The computer, among other uses, is used for general purpose computing, learning about programming, product prototyping, controlling robots, creating a media center, security systems and home automation, and as a project platform.

Setting Up Your Raspberry Pi

Obviously, before you set it up, you need to purchase your Raspberry Pi. Once you have your own Raspberry Pi, now you can go on to set it up. Luckily, setting up your Pi is pretty straightforward.

First of all, you have to make sure the Raspberry Pi operating system is installed on the SD card. You can do this using the NOOBS (new out of box software) program easily. The operating system of Raspberry Pi, known as Raspbian, and data storage are stored on a Micro SD card. This means that you can be able to set up different SD cards -each one of them booting a Raspberry Pi in different configurations. For instance, by changing the SD Card in the Pi, the device could be a robot, drone control system, camera controller, home security system, earthquake detector, weather station, radon detector, SETI cruncher, GPS, RFIDReader and many more.

You need to note that the micro SD card speed usually range up to class 10 (this is the fastest).

The class is indicated by a number in a circle. The recommended minimum useful class for Raspberry Pi is class 4. While the class 10 card will operate for a longer period of time, it (the card) tends to 'wear out' in time. Also, as with any computer data, you have to back up the SD card.

If you received your raspberry Pi pre-installed with a NOOBS SD card, you can very well skip to the Wi-Fi set up section in the book. Otherwise, you have to follow the steps below to download and install NOOBS on your SD card:

Go to this site and click the NOOBS icon. Select [Download Zip] and then unzip the folder containing the downloaded NOOBS system.

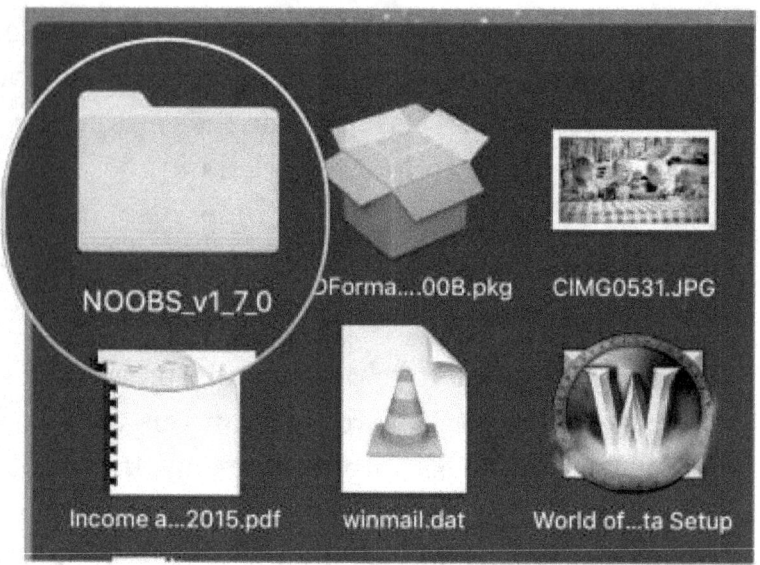

NOOBS_v1_7_0   DForma....00B.pkg   CIMG0531.JPG

Income a...2015.pdf   winmail.dat   World of...ta Setup

Follow the file labeled 'INSTRUCTIONS-README.txt' located in the unzipped NOOBS folder.

At the bottom of the desktop, you'll find the language selection window where you will choose the language of your country- for instance, in the US, select English. This will then lead to the display of the corresponding US keyboard. Now click on the Raspbian check box; click on the install icon, making sure to click 'yes' on the confirmation window. When the window labeled [OS(es) Installed] appears, just click OK.

The Wi-Fi Set Up

We will need to connect your device to the internet in order to download the programs in this book. The first step is to connect your device to a monitor and a keyboard. When you do so, login to the Raspberry Pi using these default Raspberry Pi credentials:

Username: pi
Password: raspberry

Get the network information

For this guide, we will connect to a Wi-Fi network with the information below:

- SSID (Network Name): Test Wifi Network
- PSK (Password): SecrectPassWord

Each time you see this password and network name in the guide, you have to change them to the password and network name for your local network. If you have to find the name of your local network, simply run the command below in your raspberry terminal:

sudo iwlist wlano scan

By doing so, you will get a list of all networks around you, and some important information for each network. You can however look for something that looks like <u>ESSID: Test Wifi Network</u> to find your network name.

Configure your network

You have to edit a file labeled *wpa_supplicant.conf.* to inform Pi to connect to your Wi-Fi network automatically. To open the file in nano, just type the command below:

```
Sudo nano /etc/wpa_supplicant/wpa_supplicant.conf
```

Scroll up to the end of the file, and then include the codes below to the file so that you are able to configure the network:

```
network={
  ssid="Test Wifi Network"
  psk="SecretPassWord"
}
```

Do not forget to replace it with your personal network name and password.

Press Ctrl+X followed by Y to save and close the file. Your device should now be able to connect to your network automatically. You can run the following command to check your network connection:

```
ifconfig wlan0
```

You will know you are connected when the output looks something like this:

```
wlan0  Link encap:Ethernet HWaddr 74:da:38:2b:1c:3d
       inet addr:192.168.1.216 Bcast:192.168.1.255 Mask:255.255.255.0
       inet6 addr: fe80::8727:5526:a190:b339/64 Scope:Link
       UP BROADCAST RUNNING MULTICAST MTU:1500 Metric:1
       RX packets:6917 errors:0 dropped:229 overruns:0 frame:0
       TX packets:2931 errors:0 dropped:1 overruns:0 carrier:0
       collisions:0 txqueuelen:1000
       RX bytes:10001000 (9.5 MiB) TX bytes:295067 (288.1 KiB)
```

In some instances, the device will not automatically connect; this means that you have to reboot to be able to do so. If it fails to connect after waiting 2-3 minutes, you could try rebooting the device using the command below:

```
sudo reboot
```

Initial Software Installation

Installing Samba

Samba is a program that is important to install because you will need to share files with other computers on your local network. Using Pi as a samba file server is easy and with it, you'll also be able store backups from other computers.

Samba is basically the Linux implementation of CIFS (common internet file system) /SMB (server message block) file sharing standard used by windows computers and apple computers, and supported by games consoles, media streamers and mobile applications.

In this tutorial, I am assuming you have connected a keyboard, monitor and mouse to your Pi to set up your file server. I also assume you are using an SD card that offers a reasonable storage space without needing any additional steps to make it accessible. Nonetheless, if you require more storage, you can easily mount a large external USB drive and make a Samba entry for it.

Even though we've already set up the Wi-Fi, I would still recommend you use a wired Ethernet connection for fast transfer speeds and stability –especially if you are copying over large files (but the project will still work if you use the Wi-Fi).

Setting up Samba

Samba is basically available in the standard software repositories of Raspbian. We are now going to update the repository index, ensure the operating system is fully updated and install Samba with apt-get. Open the terminal and type the following:

sudo apt-get update

sudo apt-get upgrade

sudo apt-get install samba samba-common-bin

Create your directory

We are now going to make a dedicated shared directory on the micro SD hard disk. Even though you can put it anywhere, we'll put ours at the upper level of the root file system.

sudo mkdir -m 1777 /share

The command sets the sticky part to assist in preventing the directory from getting deleted accidentally and offers everyone the permissions to write/read/execute on it.

Configure Samba to share your directory

Configure Samba to share your new directory

```
# printer drivers
[print$]
    comment = Printer Drivers
    path = /var/lib/samba/printers
    browseable = yes
    read only = yes
    guest ok = no
# Uncomment to allow remote administration of Windows print drivers.
# You may need to replace 'lpadmin' with the name of the group your
# admin users are members of.
# Please note that you also need to set appropriate Unix permissions
# to the drivers directory for these users to have write rights in it
;   write list = root, @lpadmin

[share]
comment = Pi shared folder
path = /share
browseable = yes
writeable = yes
only guest = no
create mask = 0777
directory mask = 0777
public = yes
guest ok = yes
```

Now edit the config files of Samba to make the file share visible to the PCs on the network.

sudo leafpad /etc/samba/smb.conf

In this tutorial's example, you'll have to add the entry below:

```
[share]
Comment = Pi shared folder
Path = /share
Browseable = yes
Writeable = Yes
only guest = no
create mask = 0777
directory mask = 0777
Public = yes
Guest ok = yes
```

This simply means that anyone will be able to read, write and execute files in the share by either logging in as a Samba user (we'll set up that one below) or as a guest. If you don't want to allow guest users, you can omit the line 'guest ok=yes'.

You can also use Samba to share a the home directory of a user so that they are able to access it from somewhere else on the network, or share a bigger external hard disk which lives at a fixed mount point. Simply create an entry 'smb.conf' for any path that you want to share, and it will be availed across your network upon restarting Samba.

Creating a user and starting Samba

Before you start the server, you will have to set the Samba password, which is different from your standard default raspberry password. There's however no harm in wanting to use the same, because this is just a low-security, local network project.

```
sudo smbpasswd -a pi
```

As prompted, now set a password. Lastly, try to restart Samba:

```
sudo /etc/init.d/samba restart
```

From this point henceforth, Samba will automatically start each time you power on your Pi. Disconnect the monitor, mouse and keyboard safely after making sure that you can locate the shared network folder. Let the Pi run as a headless file server.

You will also be able to locate the Raspberry Pi file server (which is by default named RASPBERRYPI) from any device on your local network. If you left the smb.conf's default settings as they are, it will show in a Windows network group known as WORKGROUP.

The next obligation we have before downloading the Ham programs is setting up the printer. Let's do that now.

Setting up the Printer

As you will realize, you will need a physical output when you are implementing the Ham radio projects and thus, having a printer on standby is a prerequisite.

Installing CUPS on your Pi and permitting remote access

To link a printer with Pi, we first have to install CUPS (Common Unix Printing System). At this point, you'll fire up your Pi then navigate to the terminal either via SSH or on the Pi itself.

You need to enter the command below at the terminal, to start installing CUPS:

```
sudo apt-get install cups
```

When the 'continue' prompt pops up, simply type Y and then tap enter. Please feel free to grab a cup of coffee because CUPS is quite a beefy install. When the base installation completes, you need to make some administrative changes. The first thing you have to do is add yourself to the user group that can access the printers or printer queue. The user group created by CUPS is known as 'lpadmin'. The Raspbian's default user (and the user that we are logged into) is 'pi'- you can however adjust the following command if you desire a different user to access the printer.

Type the following command at the terminal:

```
sudo usermod -a -G lpadmin pi
```

In case you are wondering, the switch labeled '-a' enables you to add an existing user (that is 'pi') to an existing group (that is 'lpadmin') as specified by the switch labeled '-G.'

The last part of the pre-configuration process is to enable the remote editing of CUPS configuration. You can complete the rest of the configuration through the web browser on your Raspberry Pi. If you want to, you are free to use your windows desktop browser to finish the configuration; all you will need is to toggle a bit of value in /etc/cups/cupsd.conf. Enter the command below at the terminal:

```
sudo nano /etc/cups/cupsd.conf
```

Now look for this section within the file:

```
# Only listen for connections from the local machine
Listen localhost:631
```

Now comment out the line labeled 'Listen localhost:631' and replace it with the line below:

```
# Only listen for connections from the local machine
# Listen localhost:631
Port 631
```

This will instruct CUPS to start listening for all networking interface contacts, which are directed to the port 631. Make sure to scroll down further in the config file to get to the section that is labeled 'locations'. In the following block, I have bolded any lines that need to be added to the config:

```
< Location / >
# Restrict access to the server...
Order allow,deny
Allow @local
< /Location >

< Location /admin >
# Restrict access to the admin pages...
Order allow,deny
Allow @local
< /Location >

< Location /admin/conf >
AuthType Default
Require user @SYSTEM

# Restrict access to the configuration files...
Order allow,deny
Allow @local
< /Location >
```

The 'allow @local' line addition enables the access to CUPS from any computer on your local network. Each time you do any changes in the CUPS configuration file, you will have to restart the CUPS server. Use the command below to do that:

```
sudo /etc/init.d/cups restart
```

Once you restart CUPS, you should be able to access the administration panel through any of your local network's computer by pointing its web browser at this link: http://[the Pi's IP or hostname]:631.

Add a printer to CUPS

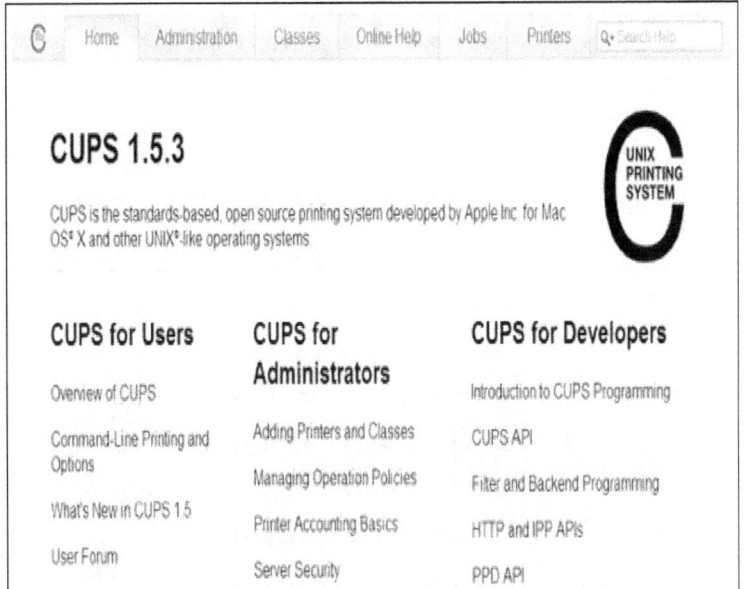

When you navigate to this link: http://[the Pi's IP or hostname]:631 ,you will be able to see the CUPS homepage, as illustrated in the image above. However, are interested in the 'administration' tab so click on it.

Next, click 'add printer' in the administration panel. If you get a warning regarding the security certificate of the site, you can ignore it by clicking 'proceed anyway'. You'll be prompted to enter a username as well as the password.

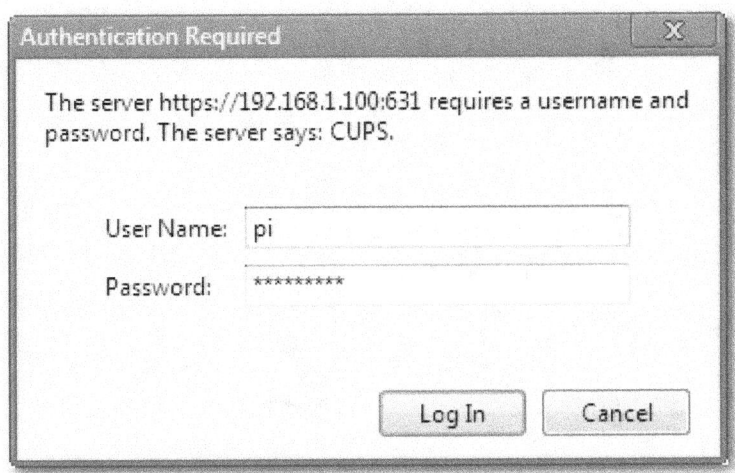

Proceed and enter the username details as well as the password of the account that you added to the 'lpadmin' group earlier on in the tutorial. For instance, if you chose to use the raspbian install (the default), the login as well as the password should be 'Pi' and 'raspberry' respectively. Now click on log in. Once you log in, you will see all the printers available (those that are local and the networked at the same time). Choose the printer you want to add to the system.

Add Printer

Local Printers: ○ HP Printer (HPLIP)
○ HP Fax (HPLIP)

Discovered Network Printers: ◉ Brother HL-2170W series (Brother HL-2170W serie

Other Network Printers: ○ AppSocket/HP JetDirect
○ Internet Printing Protocol (ipp)
○ Backend Error Handler
○ LPD/LPR Host or Printer
○ Internet Printing Protocol (https)
○ Internet Printing Protocol (ipps)
○ Internet Printing Protocol (http)
○ Windows Printer via SAMBA

[Continue]

Once you choose the printer, you'll be given the chance to edit the description, location and name of the printer, and also to enable network sharing. Since the printer we're using is already a network printer, we didn't check the 'share this printer' box.

Add Printer

Name: Brother_HL-2170W_series

(May contain any printable characters except "/", "#", and space)

Description: Brother HL-2170W series

(Human-readable description such as "HP LaserJet with Duplexer")

Location: Office

(Human-readable location such as "Lab 1")

Connection: dnssd://Brother%20HL-2170W%20series._pdl-datastream._tcp.local/

Sharing: ☐ Share This Printer

[Continue]

Once you edit the name of the printer and add a location, you will get a prompt to choose the specific driver that you want for your printer. While it discovered the printer and printer name automatically, CUPS doesn't make any attempt to choose the correct driver for you. After installing the right driver, scroll down until you are able to see a model number matching your own. As an alternative, if you have a PPD file for the printer that you've downloaded from the manufacturer, simply load it with the button labeled 'choose file'

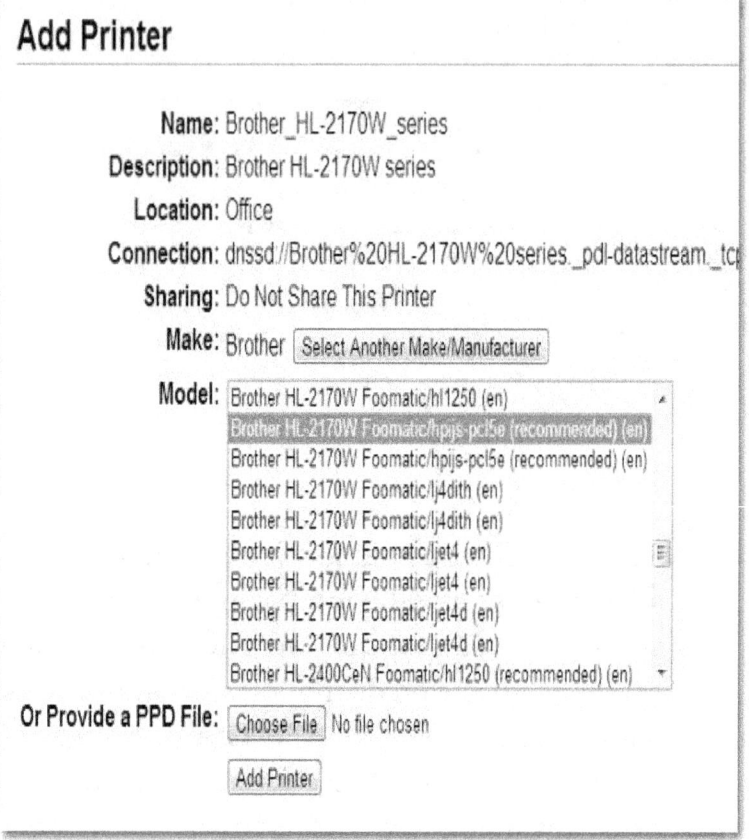

Add Printer

Name: Brother_HL-2170W_series

Description: Brother HL-2170W series

Location: Office

Connection: dnssd://Brother%20HL-2170W%20series._pdl-datastream._tc

Sharing: Do Not Share This Printer

Make: Brother [Select Another Make/Manufacturer]

Model:
Brother HL-2170W Foomatic/hl1250 (en)
Brother HL-2170W Foomatic/hpijs-pcl5e (recommended) (en)
Brother HL-2170W Foomatic/hpijs-pcl5e (recommended) (en)
Brother HL-2170W Foomatic/lj4dith (en)
Brother HL-2170W Foomatic/lj4dith (en)
Brother HL-2170W Foomatic/ljet4 (en)
Brother HL-2170W Foomatic/ljet4 (en)
Brother HL-2170W Foomatic/ljet4d (en)
Brother HL-2170W Foomatic/ljet4d (en)
Brother HL-2400CeN Foomatic/hl1250 (recommended) (en)

Or Provide a PPD File: [Choose File] No file chosen

[Add Printer]

The final step entails looking over general print settings such as the name and address of your preferred/default printer. the default paper source or size and so forth. It ought to default to the right presets, even though it never hurts to check though.

Set Default Options for Brother_HL-2170W_series

Query Printer for Default Options

General   **Printout Mode**   **Banners**   **Policies**

General

Printout Mode: Normal

Media Source: Printer Default

Page Size: Letter

Double-Sided Printing: Off

Set Default Options

Once you click the 'set default options', you'll be able to see the printer's default administration page (this is for the printer you added to the CUPS system).

Brother_HL-2170W_series (Idle, Accepting Jobs, Not Shared)

Maintenance ▾	Administration ▾

Description: Brother HL-2170W series

Location: Office

Driver: Brother HL-2170W Foomatic/hpijs-pcl5e (recommended) (grayscale, 2-sided printing)

Connection: dnssd://Brother%20HL-2170W%20series._pdl-datastream._tcp.local/

Defaults: job-sheets=none, none media=na_letter_8.5x11in sides=one-sided

Jobs

Search in Brother_HL-2170W_series: [Q▾ ....] [Search] [Clear]

[Show Completed Jobs] [Show All Jobs]

No jobs

All looks well. The real test, nonetheless, is actually getting to print something. Fire up the default text editor of Raspbian, Leafpad, and send a message:

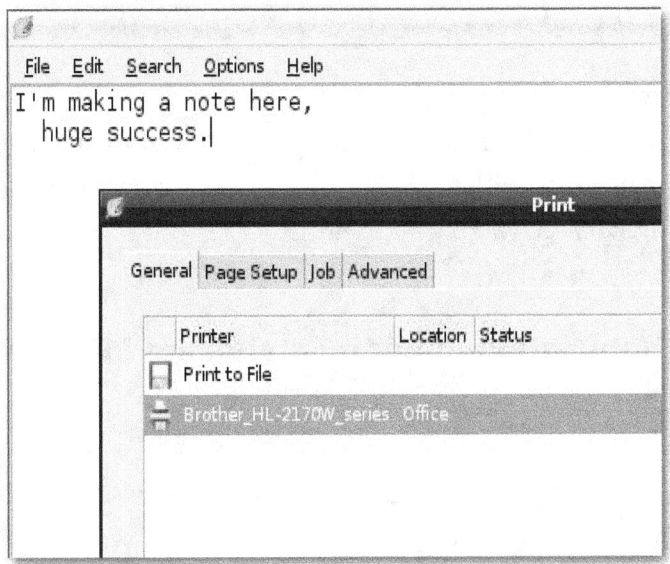

So far, if you have included the only user that has to access the printer to the 'lpadmin' group, and you have also added or included the only printer you want to access the CUPS system, that should be it. If you've got other users or additional printers you would want to add, just run through the individual steps that we've just discussed above to do so.

The Ham Radio Programs

We will now discuss the installation of important Linux amateur radio software. You can find the list of these software at http://www.raspberryconnect.com/raspbian-packages-list/item/71-raspbian-hamradio.

Hamradiomenus (it Establishes Menus)

Let's first take an overview of this program:

This program creates a sub-menu for the applications of Ham radio that you have installed. The program only appears when you have installed at least one application with a desktop entry file that contains the category entry: *HamRadio*.

The program does not have any additional hardware requirements.

How to install it

There are two ways of installing the software that you can choose from:

Start by opening a terminal session; on the taskbar, click on the terminal icon at the upper part of the screen.

There are two ways to install software. Choose one of the following methods:

Using the command line to install it

If you've not recently done an update, you can do so for new and updated Linux packages currently available with:

```
$ sudo apt-get update
```

If a new package version installed on your machine is available, the package will be upgraded with the command below. We don't have any installed packages that are ever removed by the command.

```
$ sudo apt-get upgrade
```

Tap the enter key when you get a prompt asking you if you want to continue.

Now type the command below to install the program:

```
$ sudo apt-get install hamradiomenus
```

Likewise, when you get a prompt asking you if you want to continue, tap the enter key. Once the download is complete, you have completed the installation.

Using the add or remove software menu selection

On the task bar, find the Raspberry icon and click on it. On the drop down menu that appears, click on preferences. Choose add or remove software and then enter HamRadioMenus in the search box and tap enter. Click the *hamradio menus for KDE and GNOME* selection square and click on the Apply button. Input the Pi password and when the download is complete, it means the installation is complete as well.

NOTE:

- The Hamradiomenus does not require a desktop entry.

- Configuration is also not needed.

- With regards to its operating, when a HamRadio category application is installed, the HamRadio menus icon appears

Aldo- Morse Code

Aldo is a learning tool in Morse code that provides four kinds of training methods shown as the startup menu. And they include the following:

- Blocks method- this one is used to identify different blocks that have random characters that are played in Morse code

- Koch- this is where two characters are played at top speed until you can pinpoint 90% of them, then another character is included and so forth.

- Read from file – entails sending characters that are produced from a file.

- Call sign- entails identifying random call signs played in Morse code

- Settings- setting up speed and techniques of selecting letters to be sent

- Exit program

The program does not have extra hardware requirements. For the installation however, you have two methods to choose from:

Installation by the command line

Open a terminal session by clicking the icon on the terminal on the task bar at the top of the screen. If you've not recently updated the information about the new, updated Linux packages that are available, please do so using:

```
$ sudo apt-get update
```

If a new version of a package that is installed on your machine is available, it will be upgraded with the command below. You should note that there aren't any installed packages that are ever taken away by the command.

```
$ sudo apt-get upgrade
```

In order to install the program, use:

```
$ sudo apt-get install aldo
```

When you get a prompt asking you whether you want to continue, tap the enter key. The installation will be complete when the downloading is done.

The installation by add or remove software menu selection

First click on the icon on the taskbar symbolizing Raspberry. On the drop down menu that appears, click on preferences. Now choose add or remove software and in the search box, enter aldo and tap enter. There will be a selection square labeled Morse Code training program; click on it. Click on the apply button and enter the Pi password. The installation will be complete when the download is done.

NOTE:

For the desktop entry- type the following to be able to generate a start menu window for aldo:$ sudo nano /usr/share/applications/aldo.desktop in the resulting editor:

```
[Desktop Entry]
Name=Aldo Morse Code Trainer
Comment=Amateur Radio Morse Code Trainer program
TryExec=aldo
Exec=aldo %F
Icon=aldo.png
Terminal=true
Type=Application
Categories=HamRadio;
```

Type "[Ctrl] X","Y" then enter to save the file then restart the menu by typing:

$ sudo lxpanelctl restart

For the configuration, select option 5, which refers to the setup, and then select option one for the Keyer Setup. Press the enter key twice after which the different parameters can be set. Once you complete the program setup, option 6 will take you to the main menu.

For the operating, simply click menu-HamRadio-Morse Code Training and then choose the code type you want to have generated. If you are using a HDMI HD TV as the monitor, you will hear the sound from the speaker of the monitor, or the monitor's audio output jack if the monitor doesn't have a speaker.

Chirp- configuration tool for amateur radios

Chirp is basically a tool meant for restoring, saving and management of memory, and preset data in the amateur radios. It basically supports many manufacturers and models, and also offers a way to interface with various data sources and formats. You can find the supported radio models in the site below.

http://chirp.danplanet.com/projects/chirp/wiki/Supported_Radios

For the hardware, this program will require a USB cable to connect with your device. Please see here for more information.

For the installation, you have two ways at your disposal to do that (choose one):

Installation by the command line

Open a terminal session by tapping on the terminal icon at the top of the screen (on the task bar). If you've not updated the information about the new and updated Linux packages that are available, use:

```
$ sudo apt-get update
```

If a new version of a package installed on your machine is available, the command below will be used to upgrade the package. As usual, there are no packages that have ever been erased by this command.

```
$ sudo apt-get upgrade
```

Now in order to install the program, use:

```
$ sudo apt-get install chirp
```

You will receive a prompt asking you whether you want to continue- press enter. You will know the installation is complete when the download is complete.

Installation by add or remove software menu selection

On your taskbar, simply click on the raspberry icon; on the drop down menu that appears, choose add or remove software, and in the search box, enter chirp; now press enter. You will see a selection square labeled 'configuration tool for amateur radios' click on the apply button then enter the Pi password.

You will know the installation is complete when the downloading is done.

Desktop entry file: this one is generated automatically.

Configuration: the program does not require any initial configuration.

With regards to the operation, you can use this guide.

When the download is complete, you have completed the installation.

Fldigi- the Digital Modem Program

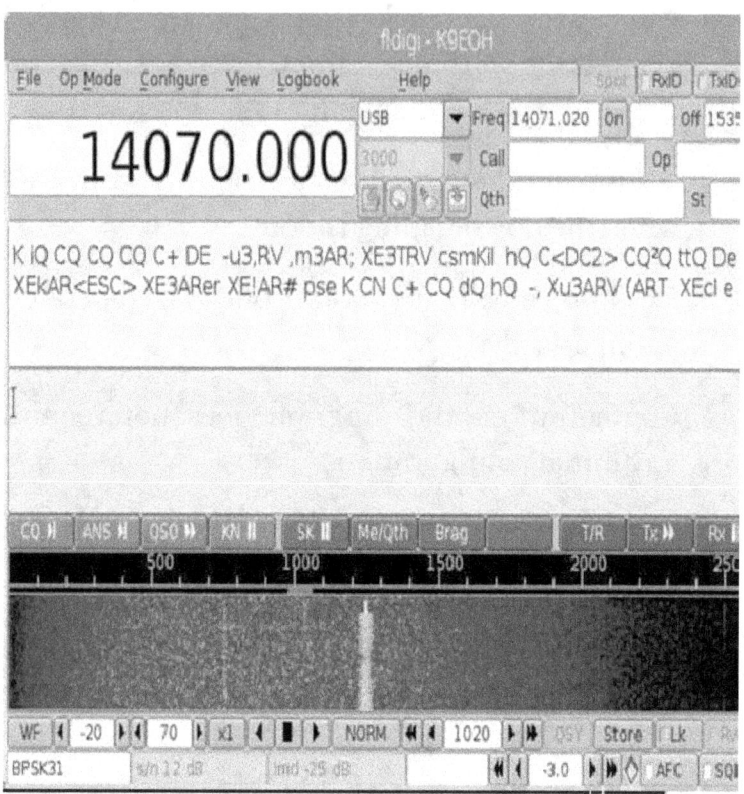

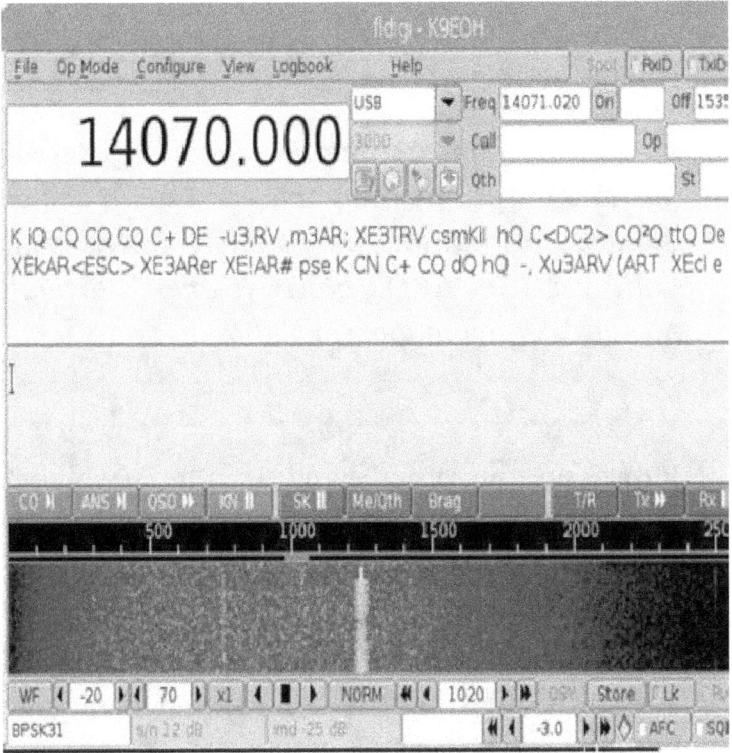

The image above describes the Fldigi screen.

Fldigi is basically a modem program that supports most current digital modes that are used by ham radio radio operators. You can also use the program to do a frequency measurement test or calibrate your sound card to WWV. This program also has a CW decoder. When you start open flarq you get a correction of errors in transmission caused by the subsequent ability to transmit as well as receive

Automatic Repeat Request (ARQ) frames. But what is Flarq?

Flarq (fast light automatic repeat request) is an ARQ specification based file transfer application that has the capacity to transmit and receive ARQ data frames through MultiaPsk on windows or Fldigi on Linux.

Flmsg on the other hand is a management editor, which is best suited for amateur radio supported standard formats of message that include MARS, ICS, HICS, NTSRadiograms, Red Cross, plain text and IARU. Its data files are ASCII that can be transmitted from one point to another via amateur radio, the internet or other electronic links.

Flwrap is a little desktop app encapsulating an image file, text file or a binary file inside a group of identifier blocks. This app is created for use to its fullest with fldigi even though you can also use it with any digital model program.

Let's now look at the hardware.

As you may have noticed, raspberry pi doesn't have an audio input capability and so you'll require an external audio adapter like the Tigertronics Signal Link USB. If you have a computer assisted transceiver (CAT), it means that the fldigi software and the transceiver will be able to pass the info to and fro. In the instance the transceiver frequency is altered, the fldigi display will also indicate and you can use it to enter the time and frequency information automatically in the built-in logging program in fldigi. A cable is needed for the CAT function (and probably the manufacturer has a cable available for you to connect the transceiver to the computer) so you should check with your transceiver's distributor to get this cable.

Nonetheless, these cables normally use RS-232 specifications. You would probably also require a RS-232 TO USB adapter for the Pi connection. Just go to Amazon.com or a similar site and enter *RS-232 TO USB adapter* and search.

NOTE/CAUTION:

The program creates a complete duty cycle with high levels of modulation when used on the air using transmitting equipment.

You thus have to refer to the manual of your transceiver and reduce the output power of your transmitter to a safe level.

For the installation, there are two ways to choose from:

Installation using the command line

Just click on the terminal icon located on the taskbar on top of your screen to open a terminal session. If you have not updated the info about new and updated packages of linux, you can do so using:

$ sudo apt-get update

(To get the available packages at the moment)

In the case that a new package version installed on your machine is available, the command below will be used to upgrade the package. There are no installed packages that are ever erased by this command though.

Installation by adding or removing the software menu selection

On the task bar, click on the Raspberry icon; on the drop down list that appears, click on preferences then choose add or remove software.

You will see a search box; enter fldigi then press enter, then click the selection squares labeled:
'digital modem program for hamradio operators', 'amateur radio file encapsulation/compression utility' and

'ham radio transceiver control program.'

Next, tap on the button labeled 'apply'; you might also be requested to enter the pi password. When the download finishes, it means the installation has finished as well.

As regards to the desktop entry file, you have to note that Fldigi and Flarq are both bundled in this package. The desktop entry files for both of them are created automatically. Nonetheless, network is included along with the Ham Radio category. In case you desire to remove the network entry, use:

```
$ sudo nano /usr/share/applications/fldigi.desktop
```

In the editor, remove network that results; from the line of categories with these results:

```
[Desktop Entry]
Name=Fldigi
GenericName=Amateur Radio Digital Modem
Comment=Amateur Radio Sound Card Communications
Exec=fldigi
Icon=fldigi
Terminal=false
Type=Application
Categories=HamRadio;
```

To save the file, press "[Ctrl] X", "Y" and press enter. repeat this for the desktop entries:

$ sudo nano /usr/share/applications/flmsg.desktop$ sudo nano /usr/share/applications/flwrap.desktop$ sudo nano /usr/share/applications/flarq.desktopUpdate the Menus:$ sudo lxpanelctl restart

As regards to the configuration, you have to note that connecting the audio interface and the transceiver control cable is important (if you are using them) before you open fldigi. The first time the program is run, you will see some 'Fldigi configuration wizard' displayed. Now tap on next.

Station

Callsign: K9EOH   Name: Jim

QTH: 5 miles (8.05 km) N /W of Spencer, Indiana, USA

Locator: EM69ni

Antenna: G5RV @ 15 feet (4.57m) running E/W

X Close   Back   Next

Now enter your call sign, Location, Name, locator or maidenhead locator

Enter your Callsign, Name, Location (QTH), Maidenhead locator and antenna info. Tap on next.

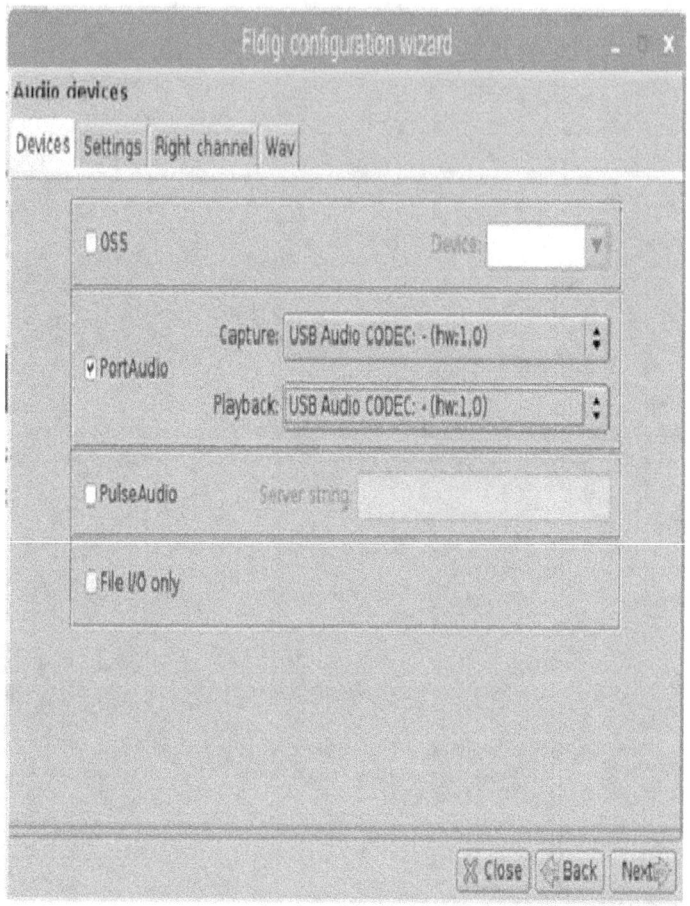

Click on the tab labeled 'devices' for SignaLink USB interface and then check the Port AudioBox; choose USB Audio CODEC: USB Audio (hw:1,0) for capture; choose USB Audio CODEC: USB Audio (hw:1,0) for Playback and select right channel tab.

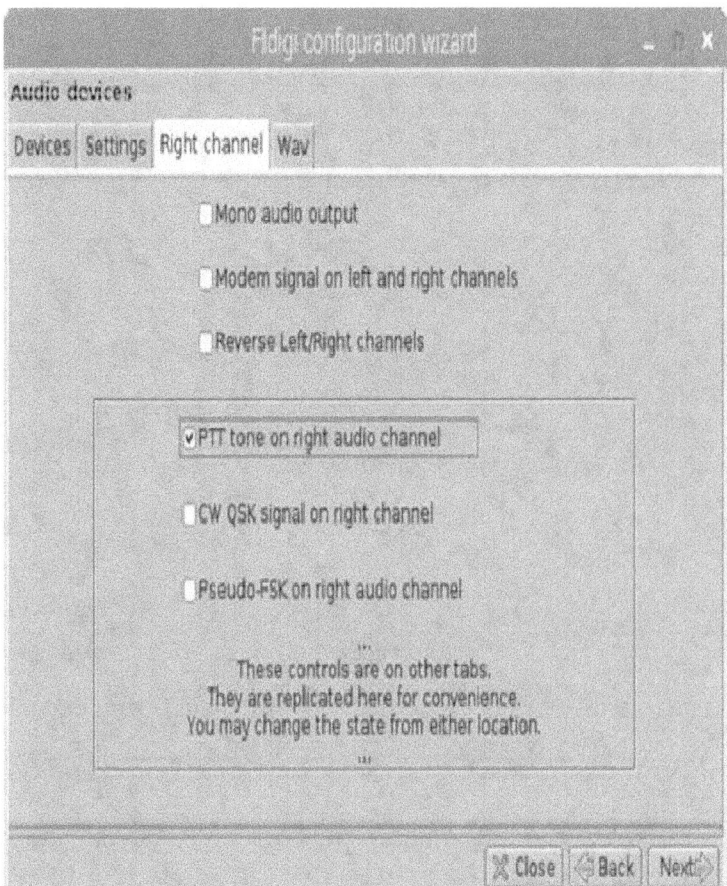

Now click the *right channel* tab and choose 'PTT tone on the right audio channel' before clicking on next. Click on the Hamlib tab in the transceiver control window.

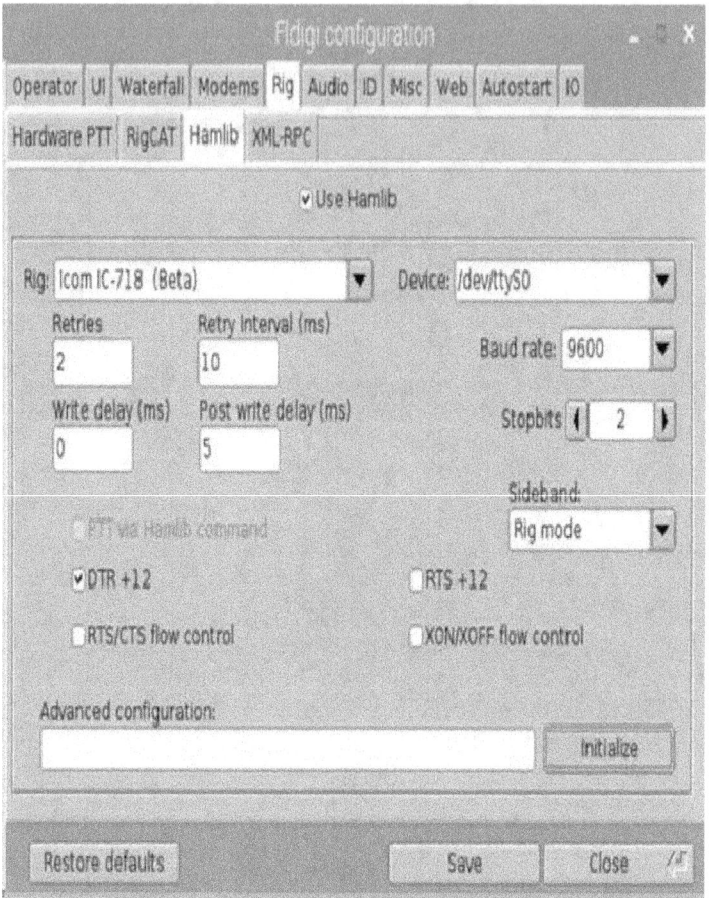

Choose the Hamlib tab and from the drop down menu that appears, choose your rig. Choose /dev/ttySO (under device- drop down menu). Next, set the Baud rate for your rig. You may need to experiment with the four boxes labeled RTS+12, DTR+12, the XON/XOFF flow control as well as the RTS/CTSflow control so that you push the Hamlib to operate your rig.

Now click on the checkbox labeled 'Hamlib' and then click the 'initialize' button as well. Tap on next until the program starts. At the top of the fldigi window, you'll find 'configure selection'; click on it too.

On the drop down window, click on Misc and then on CPU; confirm that the slow CPU box is left unchecked if you know you are using a Raspberry Pi that is either 700 MHz or faster than that. After that click on the 'NBEMS' tab- if you wish for the incoming messages to be displayed on a flmsg form, ensure you check the 'open with flmsg' checkbox. If you would want a copy to be put in a browser HTML format as well, click on the checkbox labeled 'open in browser'.

Now move to the taskbar and click on the terminal icon to open a terminal session. You can enter the command below to find the location of flmsg:

$ sudo find / -name flmsg

```
pi@HAMSHACK:~ $ sudo find / -name flmsg
/usr/share/doc/flmsg
/usr/bin/flmsg
```

As you can see from the image above, go ahead and key in the line- /usr/bin/flmsg- which does not contain /doc/ in the flmsg: window. Click one save and close then click on the 'x' in the upper right corner to restart the program. You will need to click on the button labeled 'yes' as well to confirm quit.

After that, choose fldigi through the menu system in the task bar at the top of your screen (raspberry icon), Fldigi and HamRadio.

As regards to the operating, the digital brands include the following:

1.805-1.838, 3.522-3.620, 7.025-7080, 10,137-10.142,14.070-14.107, 18.098-18.106, 21.070-21.540, 24.197-24.922 and 28.076-28.120 MHz.

You can get the full list along with the modes here:

Gpredict – For Satellite Tracking

Gpredict is an application for real-time orbit prediction and satellite tracking. This application can track numerous number of satellites and show their position and related data in tables, lists, maps and if course polar plots.

The app can also be able to make predictions of a satellite's time for future passes and offers you all the information about all passes.

Gpredict has been seen as being a unique satellite tracking program because it enables you to assemble the satellites in terms of visualizations modules, and you can configure each one of them (the modules) independently from the others, thus offering you unlimited flexibility with regards to the look and feel of the modules. Naturally, Gpredict also allows you to track satellites comparatively to various observer locations all at the same time.

The installation and usage

Start the terminal window and type the following   command:

```
sudo apt-get install gpredict
```

Your Raspberry device will then download and install the software for you and take you back to the command prompt when it completes.

To start up the software, just type the following from your terminal window:

```
gpredict
```

With that, the software now starts up; you have to perform a bit of configuration to let the program to know your current location (Copenhagen- is the default).

Click edit or preferences and choose the ground stations tab. Then press add new and proceed to add your location details, altitude, longitude and latitude. Once you save that, you can now delete the sample location of Copenhagen and turn your own into the default.

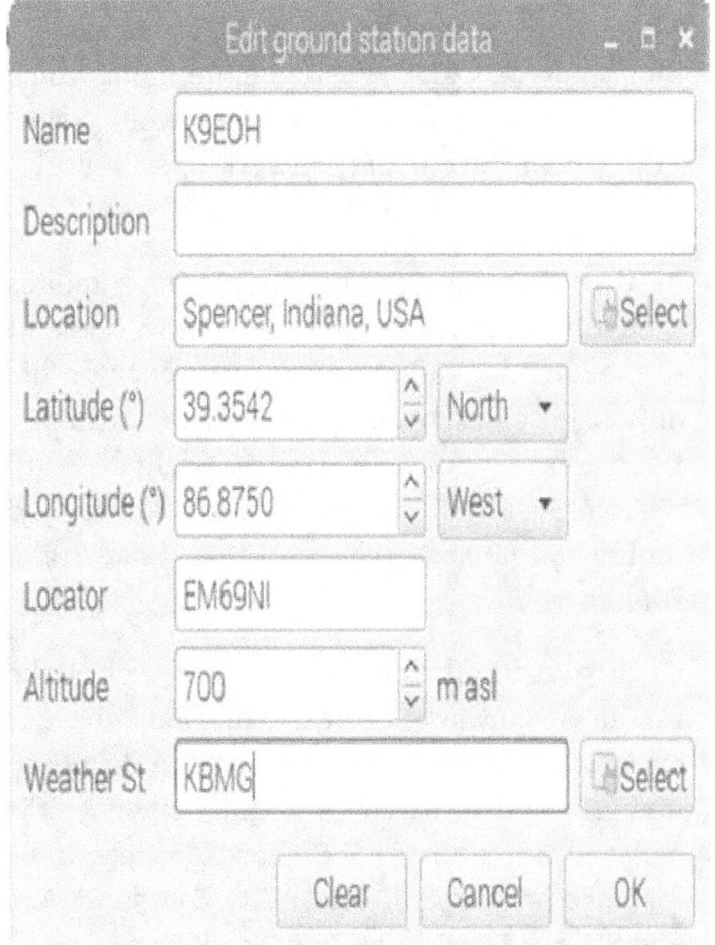

While still in preferences, you can choose your screen layout. If you have a large monitor, then the optimum is table, Map, Polar and a wide single sat (you however have the freedom to play around and see the one that suits you best). You may have to restart the program for the changes to take effect.

About now, it might be a good idea to check that you've got the latest satellite data, which the program can download for you. Select Edit/Update TLE from network and then give it a moment for the update to finish.

Lastly, you will want to configure the satellites that you are tracking. The software will default to the amateur radio module. Just click on the options or shortcuts on the module –at the top right area of the main window, right beneath the main windows control: close, maximize and minimize. Just click on that and then configure.

Once you do all that, you can now select the satellites you want to track. While Fancube-1 or even AO-73 is not being displayed under those names, you can search for it and include it- it is believed to be 2013-066B though. I personally included other satellites like ISS, VO-52, FO-29 and SO-50. Your choice may be different.

At this point, you should be able to see the location of all your satellites as plotted on the screen.

If you want more details about a specific satellite, simply highlight it in the list at the bottom and right click to choose 'future passes' or 'show next pass.'

I can say that G-predict does work well on the Pi. Even though it doesn't appear to max the processor out really well, at least it is quick to start up and shut down- if you want to engage yourself with something else.

Twclock- world clock

As a ham operator, the clock program will prove to be very helpful. Apart from displaying GMT and local time, it is able to display the current time of all the major cities around the globe. It also has an alarm to inform you when the time for a station ID arrives.

As you will realize, you can set the ID alarm to a preferred delay of seconds and minutes. The alarm will inform you that it's time to ID or all of the ways below:

- Beep the PC's speaker

- Blink the alarm button

- Send a call in CW to you using pulseaudio via your sound card

You could feed the CW audio to your rig to allow transmission of the ID is automatically. The CW is produced using code from qrq.

You will also find an auto reset choice. This choice starts the next time out automatically without any action from you, the user.

So, you need to connect your soundcard's output to the 'audio-in' pin of your rig's accessory jack so that it is transmitted.

Essentially, this program is essentially a clock that is tailored for the ham radio operators, those who want to know the time in some other place in the world, or just about anyone who has had enough of the same look of the ordinary clock. The program shows the current date and time in different cities around the world at the same time.

The installation

The installation process is simple, and it uses the CLI (command line interface).

Begin by updating the repository index on your Raspberry Pi by using:

```
Sudo apt-get update
```

Now search for the Raspberry Pi repository index for the twclock programs via:

```
Apt-cache search twclock
```

Next, install the program:

```
Sudo apt-get install twclock
```

When that is done, start twclock. In the graphic user interface of your computer, open the top left application menu and choose accessories, twclock. You can open two programs of twclock at the same time- one of them set to local time and the other one to GMT time.

Uninstalling it is also simple:

```
sudo apt-get uninstall twclock
```

So far, you have pretty much enough programs to get you started with amateur radio activities. However, since you have to be really good at this, I will add a couple more, under the next chapter in which we will look at full Ham projects.

More Ham Radio Programs, And Projects

GNU Radio Companion

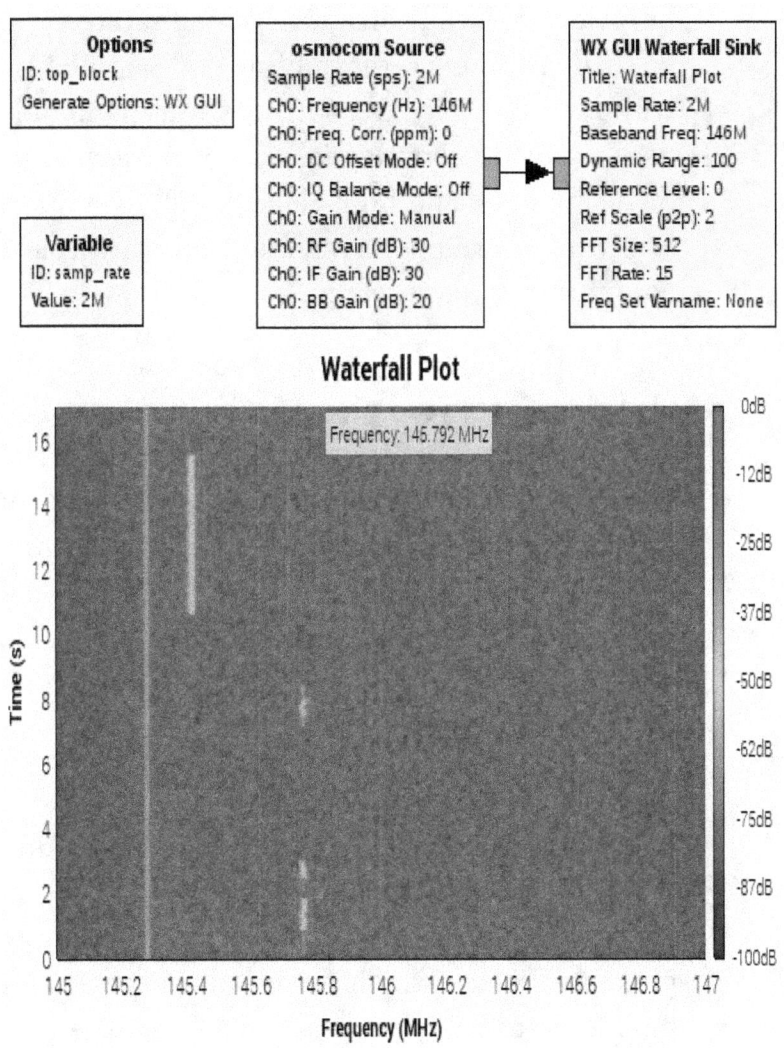

GNU Radio is basically a free software development toolkit that essentially offers blocks to implement software radios. You can use it with readily available cheap external RF hardware to build software defined radios or in an environment resembling a simulation without any hardware. This software is mainly used by commercial and academic environments, and hobbyists as well to support real-world radio systems and wireless communications research.

The applications of GNU radio are mainly written with Python programming language, and the supplied performance-critical path of signal processing is usually implemented in C++ with, where available, processor floating point extensions. This means that the developer is capable of implementing radio systems, which are real time and high throughput in a rapid-application development environment that is simple to use.

Even though GNU is not primarily a simulation tool, it supports development of signal processing algorithms with pre-recorded or generated data- which avoids the necessity of having actual RF hardware. The package also contains a graphical tool known as gnuradio-companion, which creates signal flow graphs and generates flow-graph source code as well. We also have many different tools and a utility program included.

Why do you need GNU?

GNU Radio does all the signal processing; you can write applications with it to get data out or even push data into the digital streams, which is then transmitted via hardware. A GNU radio contains channel codes, filters, synchronization elements, vocoders, demodulators, equalizers and other elements (we usually refer to these as blocks in the GNU jargon) that are usually located in radio systems. More importantly, it also has a way of connecting blocks and can manage the way data usually passes from one block to another. It is actually very easy to extend a GNU Radio: if you get a particular missing block, you can create it and add it quickly.

Moving on;

GRC (GNU Radio companion) is a graphical user interface that is used in the development of GNU Radio applications. As you can see in the images above, you can use some inexpensive SDR USB dongles (>$20) such as the NooElec R820T SDR & DVB-T NESDR Mini (more on this later).

You can use the process below to install it using the command line interface:

Update the raspberry Pi repository index using:

Sudo apt-get update

Now make a search in the Raspberry Pi repository index for Gpredict programs with:

Apt-cache search gnuradio

- pi@raspberrypi:~ $ apt-cache search gnuradio
- gnuradio - GNU Radio Software Radio Toolkit
- gnuradio-dev - GNU Software Defined Radio toolkit development
- gnuradio-doc - GNU Software Defined Radio toolkit documentation
- gr-air-modes - Gnuradio Mode-S/ADS-B radio
- gr-fcdproplus - Funcube Dongle Pro Plus controller for GNU Radio
- gr-osmosdr - Gnuradio blocks from the OsmoSDR project
- libair-modes0 - Gnuradio Mode-S/ADS-B radio
- libgnuradio-analog3.7.5 - gnuradio analog functions
- libgnuradio-atsc3.7.5 - gnuradio atsc functions
- libgnuradio-audio3.7.5 - gnuradio audio functions
- libgnuradio-blocks3.7.5 - gnuradio blocks functions
- libgnuradio-channels3.7.5 - gnuradio channels functions
- libgnuradio-comedi3.7.5 - gnuradio comedi instrument control functions
- libgnuradio-digital3.7.5 - gnuradio digital communications functions
- libgnuradio-dtv3.7.5 - gnuradio digital TV signal processing blocks
- libgnuradio-fcd3.7.5 - gnuradio FunCube Dongle support
- libgnuradio-fcdproplus0 - Funcube Dongle Pro Plus controller for GNU Radio
- libgnuradio-fec3.7.5 - gnuradio forward error correction support
- libgnuradio-fft3.7.5 - gnuradio fast Fourier transform functions
- libgnuradio-filter3.7.5 - gnuradio filter functions
- libgnuradio-iqbalance0 - GNU Radio Blind IQ imbalance estimator and correction

Install gnuradio with:

Sudo apt-get install gnuradio

Next, install gr-osmosdr with:

Sudo apt-get install gr-osmosdr

Install gr-air-modes with:

Sudo apt-get install gr-air-modes

Now start GRC- In your desktop's GUI, select programing, GRC (after opening the top left application). Make a GRC 2 meter (146 MHz) waterfall plot with a SDR.

Plug in the device below to the Pi: (you can find more details about it from NooElec).

I personally used a USB hub for the device because it is too large and would thus block the rest of the USB ports.

Now drag the RTL-SDR Source (in sources) to the left work window. Drag the WX GUI Waterfall sink (in instruments, WX) to the left work window.

Now wire the source output of osmacom to WX GUI Waterfall Sink.

In the variable and ID same_rate, alter the value to 2 MS/s

In the source output of osmacom to the WX GUI Waterfall sink, alter the frequency to the values: 146E6.

In the top icon menu, choose 'generate the flow graph.'

Save the file in the file menu.

Finally, choose execute the flow graph in the top icon menu.

The set up

It always seems quite amazing to me what a humble USB TV dongle can do- considering you can pick up one for about £10 together with the open source SDR software. As you can see in the picture above, it is very simple to plug one of them in one of the USB ports, with the supplied antenna fixed. Please visit this page for more information about RTL-SDR.

Now that we are re-purposing a TV tuner which the Linux kernel supports, and which it would otherwise claim and for television reception, we have to first of all make the kernel to stop doing so. We will edit the '**/etc/modprobe.d/raspi-blacklist.conf**' file and proceed to add the following line below:

blacklist dvb_usb_rtl28xxu

Install Software (RTL-SDR)and the GNU Radio support as follows:

$ sudo apt-get install rtl-sdr gr-osmosdr

You will also require setting up a new udev rule so that you are able to access the device as a non-root user- you however first have to ascertain the USB ID. Make sure the tuner is plugged in and then type the following:

$ lsusb

This should give you the following:

Bus 001 Device 004: ID 0bda:2832 Realtek Semiconductor Corp. RTL2832U DVB-T

Now you need to create the /etc/udev/rules.d/20.rtlsdr.rules file with the following line:

SUBSYSTEM=="usb", ATTRS{idVendor}=="**obda**", ATTRS{idProduct}=="**2832**", GROUP="adm", MODE="0666", SYMLINK+="rtl_sdr"

At this point, you can restart Udev; nonetheless, since you also had a kernel module blacklisted, you probably should consider rebooting as the easiest way.

The test

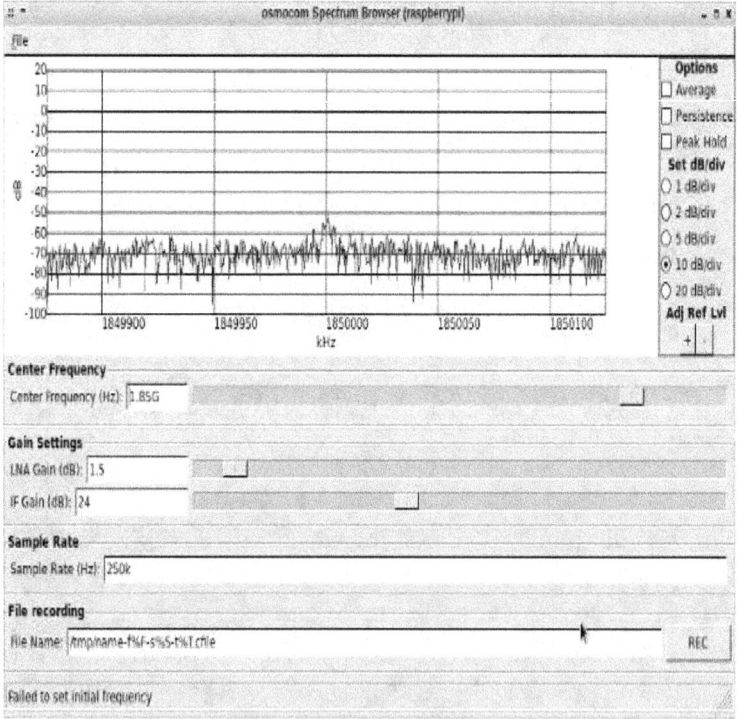

You can run the FTT application to get a modest spectrum display, which is offered as part of the gr-osmocom software.

```
$ osmocom_fft
```

If you then look at the CPU load, you can see that you've got a lot of capacity to spare, with just a single core at about 70% utilization.

```
top - 14:21:22 up 14 min,  3 users,  load average: 1.08, 0.69, 0.40
Tasks:  83 total,   2 running,  81 sleeping,    0 stopped,    0 zombie
%Cpu(s): 17.2 us,  1.3 sy,  0.0 ni, 81.3 id,  0.0 wa,  0.0 hi,  0.2 si,  0.0 st
KiB Mem:    764060 total,   203720 used,   560340 free,    17100 buffers
KiB Swap:   102396 total,        0 used,   102396 free,   102792 cached

  PID USER      PR  NI  VIRT  RES  SHR S  %CPU %MEM    TIME+  COMMAND
 2189 pi        20   0  286m  99m  53m R  70.4 13.4  3:54.14 osmocom_fft
 2103 pi        20   0 10176 3488 2856 S   3.3  0.5  0:22.10 sshd
 2185 pi        20   0  4948 2464 2100 R   0.7  0.3  0:03.64 top
   43 root      20   0     0    0    0 S   0.3  0.0  0:00.54 kworker/0:1
    1 root      20   0  2248 1332 1232 S   0.0  0.2  0:01.69 init
```

The gr-air-modes

Over the recent past, so much has been written about how to use the RTL-SDR hardware along with the GNU Radio-based gr-air-modes software, to get position and the heading information from the aircraft Mode-S transponders. While many writers state that they tried using a Raspberry Pi Model B (and a laptop)- this hasn't proven to have enough processing power and the result is usually buffer underruns.

A couple more dependencies are needed, in order to build gr-air-modes.

```
$ sudo apt-get install cmake libboost-dev sqlite pyqt4-dev-tools liblog4cpp5-dev swig
```

Having installed these ones, the sources can be cloned from GitHub as follows:

$ git clone Github

To then create and install:

```
$ cd gr-air-modes
$ mkdir build
$ cd build
$ cmake ../
$ make
$ sudo make install
$ sudo ldconfig
```

You can now use the following to run the application:

$ modes_rx -s osmocom

And with only a small antenna and a significant number of miles from the closest airport, you will still manage to receive no shortage of output.

```
pi@raspberrypi ~ $
pi@raspberrypi ~ $ modes_rx -s osmocom
linux; GNU C++ version 4.9.1; Boost_105500; UHD_003.007.003-0-unknown

gr-osmosdr 0.1.3 (0.1.3) gnuradio 3.7.5
built-in source types: file osmosdr fcd rtl rtl_tcp uhd miri hackrf bladerf rfspace airspy
Using device #0 Realtek RTL2832U SN: 00000001
Found Elonics E4000 tuner
Invalid sample rate: 4000000 Hz
Gain is 34
Rate is 4000000
Using Volk machine: generic_orc
(-40 0.00000000) No handler for message type 24 from 5b44c1
(-36 0.00000000) Type 4 (short surveillance altitude reply) from 5125f at 87700ft (GROUND ALERT)
(-36 0.00000000) Type 0 (short A-A surveillance) from 648693 at 200ft (speed <75kt)
(-37 0.00000000) Type 4 (short surveillance altitude reply) from 4ee766 at 38300ft (AIRBORNE ALERT)
(-39 0.00000000) Type 4 (short surveillance altitude reply) from 4d3bf at 125500ft (AIRBORNE ALERT)
(-41 0.00000000) No handler for message type 24 from 4f5d0d
(-36 0.00000000) Type 0 (short A-A surveillance) from 4d6e4a at 12400ft (No TCAS)
(-38 0.00000000) Type 5 (short surveillance ident reply) from 51ebf with ident 242 (AIRBORNE ALERT)
(-35 0.00000000) Type 5 (short surveillance ident reply) from a1bd2c with ident 134 (aircraft is on the ground)
(-35 0.00000000) Type 4 (short surveillance altitude reply) from 330fce at 3425ft (SPI ALERT)
(-38 0.00000000) Type 0 (short A-A surveillance) from a54ae6 at 31500ft (speed 75-150kt)
(-37 0.00000000) Type 5 (short surveillance ident reply) from 9fcfc1 with ident 1300 (AIRBORNE ALERT)
(-40 0.00000000) Type 4 (short surveillance altitude reply) from c966c8 at 1700ft (SPI)
(-40 0.00000000) Type 0 (short A-A surveillance) from 43c8bd at 53400ft (No TCAS)
(-41 0.00000000) No handler for message type 24 from c02d66
(-40 0.00000000) Type 5 (short surveillance ident reply) from a79fe0 with ident 6108 (SPI)
(-35 0.00000000) Type 0 (short A-A surveillance) from ae43d7 at 98800ft (TCAS resolution inhibited)
```

Once again, not to mention, with a lot of headroom to spare:

```
top - 09:28:22 up 18:23,  3 users,  load average: 0.50, 0.37, 0.19
Tasks:  83 total,   1 running,  82 sleeping,   0 stopped,   0 zombie
%Cpu(s):  5.4 us,  1.2 sy,  0.0 ni, 92.4 id,  0.9 wa,  0.0 hi,  0.0 si,  0.0 st
KiB Mem:    764060 total,   318912 used,   445148 free,    27420 buffers
KiB Swap:   102396 total,        0 used,   102396 free,   207648 cached

  PID USER      PR  NI  VIRT  RES  SHR S  %CPU %MEM    TIME+ COMMAND
 9067 pi        20   0  282m  80m  40m S  25.2 10.7  0:08.57 modes_rx
 9086 pi        20   0  4948 2524 2156 R   0.7  0.3  0:00.14 top
   57 root      20   0     0    0    0 S   0.3  0.0  0:00.25 kworker/2:1
    1 root      20   0  2556 1364 1256 S   0.0  0.2  0:02.40 init
    2 root      20   0     0    0    0 S   0.0  0.0  0:00.00 kthreadd
    3 root      20   0     0    0    0 S   0.0  0.0  0:00.09 ksoftirqd/0
    5 root       0 -20     0    0    0 S   0.0  0.0  0:00.00 kworker/0:0H
```

Build a FlightAware PiAware Ground Station

You can construct and actually run your ADS-B ground station which you can install anywhere and actually receive data in real time on your computer directly from airplanes. Your ground station will be able to run FlightAware's PiAware software to be able to track any flights that are within a 100-300 mile radius (this is the line of sight- the range depends on the antenna installation) and will feed data to FlightAware automatically. You can track flights off your PiAware gadget directly or through FlightAware.com.

NOTE: As a token of appreciation from FlightAware, you will receive the following for sending ADS-B data:

- Live data through flightaware.com- this is subject to a delay of standard data processing of up to two minutes.

- An access to live data (that is up-to-the-second) that is received by the local device (this is accessible with a local network connection from the stats page).

- Data from the local device highlighted on the track logs of FlightAware.

- Detailed site performance statistics.

- A free <u>enterprise account</u> for just $89.95 per month

Getting started

The process is quick and very easy. With the instructions below, you'll find the process short- it actually takes about 2 hours to complete the project- the parts themselves cost about $100.

Apart from your Pi, and other essentials we've already discussed like an SD card, you will need the following:

USB SDR ADS-B (Automatic Dependent Surveillance-Broadcast) Receiver (Pro Stick Plus or FlightAware recommended)

The USB SDR ADS-B receiver will transform the 1090 MHz radio signal to a form that your computer can actually understand

TIP: if you are selecting between the Pro Stick Plus and FlightAware Pro Stick, remember that the Plus contains an on-board filter which works very well in places that contain a lot of radio noise- like urban environments.

1090 MHz Antenna

To start, you can buy an indoor antenna. If you are using the FlightAware USB adapter, ensure the antenna contains an SMA connector.

If you are using a telescoping mast antenna, make sure to break it down to a quarter wave length of 1090 MHz -6.9 cm- so that you maximize reception.

Install PiAware on your SD card

I will now show you how to install the software on your SD card if you are using a windows or Mac OS; however, I have to remind you to ensure you select the right drive to install the image.

1. Windows

Download the PiAware software on Raspbian Linux ZIP whose size is 310 MB then save the file on your computer. By clicking the link, it downloads the file into your computer's download folder automatically.

Next, visit https://etcher.io/ to download the SD card writer (which is about 80MB) in size and follow the directions to install the program.

You can find the info about your windows OS under Control Panel >all control panel items> system. To see all the control panel options, view the control panel with little icons.

Next, open Etcher. You may have to run Etcher as administrator. To do so, right click \run as administrator\

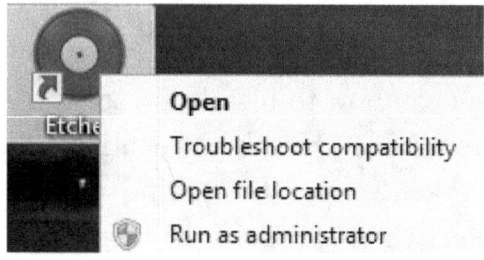

Choose the PiAware zip file

Choose the correct Micro SD card USB drive carefully. This process usually overwrites any data that's on the drive so be careful with it to ensure you don't lose data accidentally. For instance, you can consider ejecting the external hard drives as well as any other storage devices to eliminate any likelihood of overwriting the data by accident.

On 'my computer' screen, confirm the drive. In the picture below, the removable disk (F) is the micro SD card:

When you are sure that the correct drive is selected, click 'flash!'

Once the PiAware has been installed, Etcher should actually automatically eject your SD card automatically; otherwise, you should manually eject your SD card then remove it from your computer if this doesn't happen.

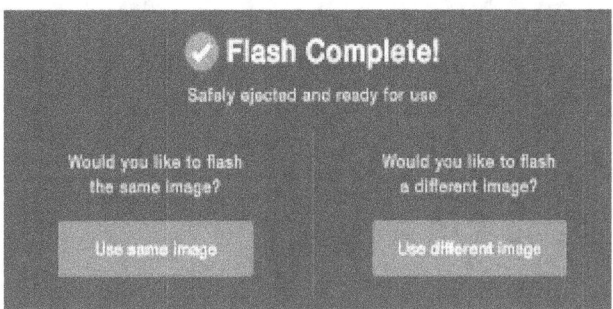

2. Mac OS X

Start by downloading PiAware on Raspbian ZIP, which is 310 MBs large. Save the file on your computer. Like I mentioned earlier, the file will download into your 'downloads' folder once you click on the link.

Next, visit https://etcher.io/ to download the SD card writer, which is about 80 MB in size and while following the directions indicated, install the program.

Now open etcher; you may have to run the program as administrator (use control + open application)

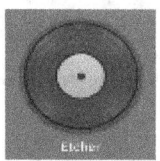

Now select the PiAware zip file and carefully choose the right Micro SD card USB drive carefully as you run the risk of overwriting important data. To avoid overwriting the data accidentally, simply remove any connected external hard drives.

On the 'finder' screen, confirm the drive. As you can see in the image below, the micro SD card is shown as 'NO NAME'.

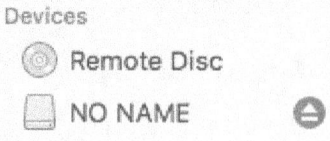

You will need administrative privileges on the computer- if a prompt appears, sign into an administrative account.

When you are sure the right drive is selected, press 'Flash!'

Once PiAware is installed, Etcher should eject the SD card automatically; otherwise, you can manually eject the card and remove it from your computer.

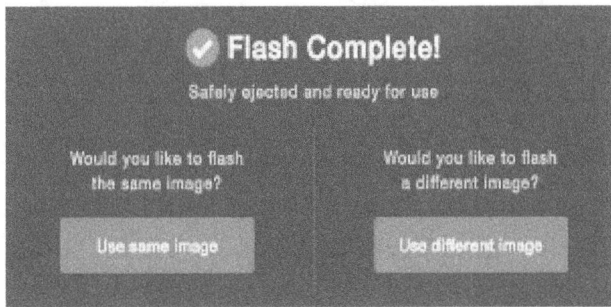

Start your PiAware device

Slide the SD card containing PiAware into your Pi.

You can also put the case on the Pi- this is optional though.

Plug the USB SDR ADS-B receiver into your Pi.

Plug the antenna cable into the USB SDR ADS-B dongle tightly.

If you are not using Wi-Fi, plug the internet cable (Ethernet) in (otherwise, you can skip this part).

Plug the power into your Pi.

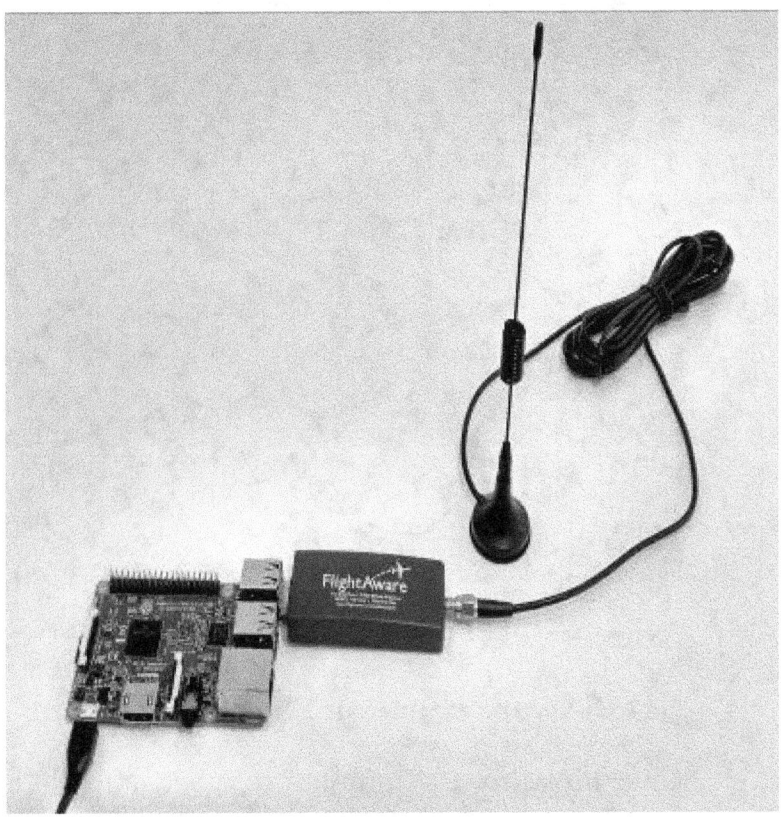

Make sure you see a solid red LED and also a blinking green LED on your Pi as well green and yellow LEDs next to the Ethernet jack.

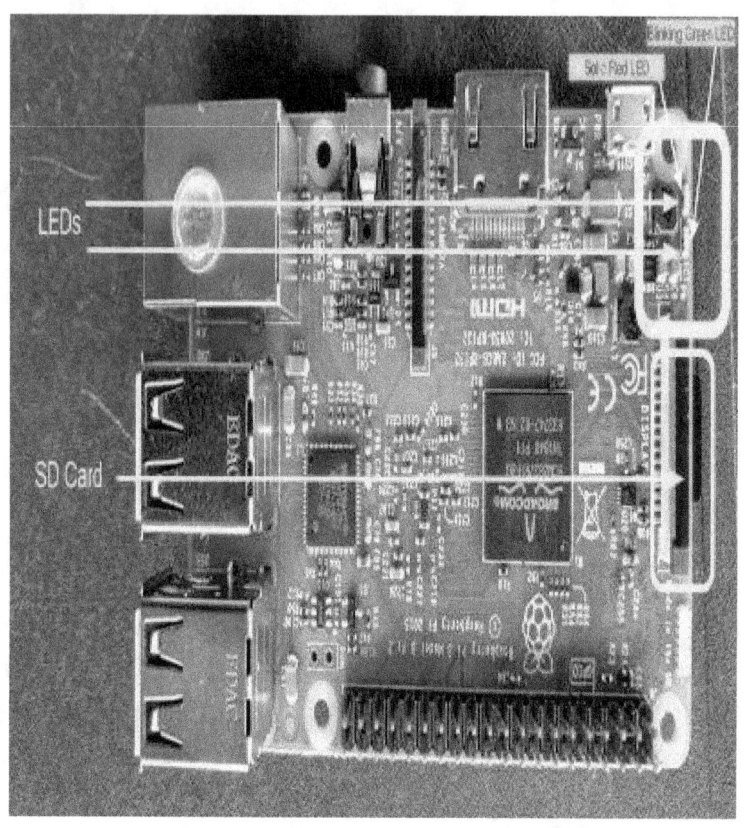

Get PiAware client on FlightAware.com

Give PiAware 4-5 minutes to start, and then you can link your FlightAware account with the PiAware device in order to get all the benefits.

When your device begins running, do the following:

Look up the IP address in your router admin and then go to the assigned IP address on the same network in a browser. If the device has not been claimed, you will get a display of the link to claim PiAware device.

Alternatively visit FlightAware.com to get your PiAware client.

If after five minutes your device has not been displayed as claimed, you can try to restart the device; if that does not work, you can reconfirm the accuracy of Wi-Fi settings (if you are using Wi-Fi).

View your ADS-B statistics

Visit https://uk.flightaware.com/adsb/stats to see your ADS-B stats.

FlightAware will start to process your data as immediately and display your statistics in 30 minutes.

Click on the gear icon situated to the right of the Site name to configure your antenna height and location on your statistics page. Multilateration, which is also known as MLAT, works by pinpointing the aircraft location by knowing the site locations that got messages from the aircraft.

The stats page of FlightAware also tells you about your device's local IP and offers a link for direct connection. This is where you can get a link to SkyView- this is a web portal for observing flights that the receiver is getting messages from on a map.

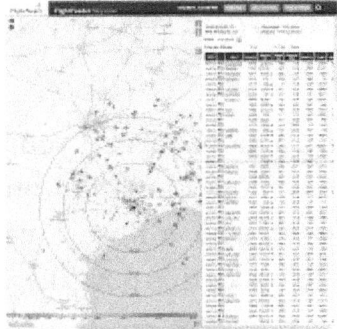

Congratulations!

Don't forget that the signals from the aircraft are not made to go through objects; this means that you should ensure the antenna is located in 'line of sight' to the sky without any obstructions. As you may realize, the most optimum installations, which are usually installed on a roof outdoors, contain a range of over 400km/250 mi.

Now that you are done, PiAware is yours to enjoy, and so is FlightAware, which comes with extra features that are exclusively reserved for feeders of ADS-B like you.

Remote Ham Radio Operation Via Raspberry Pi

In this project, you'll learn how to use your Raspberry Pi to set up a remote operation on your ham radio. In this case, you will use your Wi-Fi network, and VPN to make this possible.

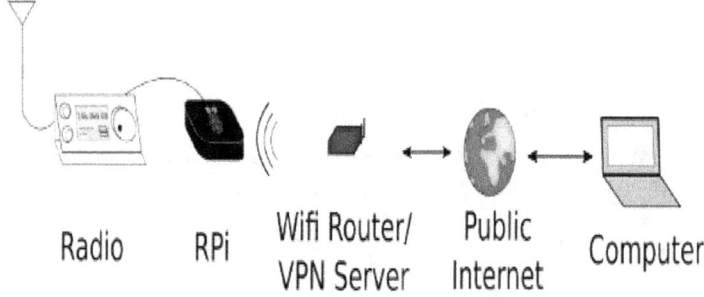

| Radio | RPi | Wifi Router/ VPN Server | Public Internet | Computer |

Why this is important

If you are a person who travels a lot and want to still continue using your rig even when you are away, this will prove important. Also, if you are a person who wants to set up your radio as well as the antenna away from where you live e.g. you are an apartment dweller and have parents or friends who stay in the boonies and would allow you to erect an antenna.

You will require the following to complete this project:

- ICOM-7100 all-mode radio

- Buffalo DD-WRT router/VPN server

- Raspberry Pi B+,

- Laptop that runs Ubuntu 15.04

NOTE: In this post, we'll focus on radio control and networked audio but not so much about setting up the VPN.

The basics

For this project, we want our raspberry to remain a 'headless' remote server. This means we don't need a keyboard or screen. You thus have to be able to SSH from your client into it. This is simple. You can set up private/public keys so that you don't have to type a password each time you log in. Just follow the link to set that up.

Build a PulseAudio 6.0 on your Pi

You have to use the PulseAudio Linux sound system in order to pipe the audio over the network. PulseAudio 2.0 occurs in the Raspberry Pi repositories. You first need to install some build dependencies:

sudo apt-get install -y libltdl-dev libsamplerate0-dev libsndfile1-dev libglib2.0-dev libasound2-dev libavahi-client-dev libspeexdsp-dev liborc-0.4-dev libbluetooth-dev intltool libtdb-dev libssl-dev libudev-dev libjson0-dev bluez-firmware bluez-utils libbluetooth-dev bluez-alsa libsbc-dev libcap-dev checkinstall

There is a libjson-c dependency you can build yourself:

git clone git://github.com/json-c/json-c.git

cd json-c

./autogen.sh

./configure

make

sudo make install

The clean pulse files that you may need to install:

(You need to note that this may break the other installed packages such as pygame).

```
sudo apt-get remove libpulse0
```

Next, acquire the PulseAudio source code; compile it. You can use checkinstall so that you can remove the built package easily later on.

git clone git://anongit.freedesktop.org/pulseaudio/pulseaudio

cd pulseaudio

./bootstrap.sh

./configure --prefix=/usr --sysconfdir=/etc --localstatedir=/var --disable-bluez4 --disable-rpath --with-module-dir=/usr/lib/pulse/modules

make

sudo checkinstall --pkgversion 6.0 --fstrans=no

sudo addgroup --system pulse

sudo adduser --system --ingroup pulse --home /var/run/pulse pulse

sudo addgroup --system pulse-access

sudo adduser pulse audio

sudo adduser root pulse-access

After that, you have to build the init.d script in order to get pulseaudio to start automatically in system mode- while this is not recommended, it makes a lot of sense on Pi. Proceed to Reboot to start PulseAudio or simply enter:

sudo service pulseaudio start

The configuration of PulseAudio to pass sound through the network

On Pi, simply edit the file /etc/pulse/system.pa, uncomment or include a line in order to turn on the module-native-protocol-tcp, to allow the local network to communicate with it.

Enable networked audio

load-module module-native-protocol-tcp auth-ip-acl=127.0.0.1;192.168.0.0/16

You may also find it great to comment out the module-suspend-on-idle line in it because it produced 10 seconds lags at first. Simply plug in the radio into your Pi and then proceed to restart PulseAudio with the following:

sudo service pulseaudio restart

Configure PulseAudio on the client

PulseAudio runs on a basis of 'per-user' on Ubuntu; not in system mode. In this case therefore, we can turn off and on the modules with scripts and that's what you'll do. Create a script and put in it commands like so:

```
#!/bin/sh


pactl load-module module-tunnel-source server=raspberrypi
source_name=icom_source

pactl load-module module-tunnel-sink server=raspberrypi
sink_name=icom_sink



# radio -> laptop speakers

pactl load-module module-loopback source=icom_source


# laptop microphone -> radio

pactl load-module module-loopback sink=icom_sink
source=alsa_input.usb-0d8c_C-Media_USB_Audio_Device-00-
Device.analog-mono
```

You need to adjust your Pi's hostname (or simply directly type its IP address in). You also have to adjust the second loopback module source. It is set to the USB microphone; but the microphone of your laptop will certainly be different. Now use the command below to obtain a list of sources on your computer and pick the name of the one you desire.

```
pacmd list-sources
```

Make your script executable by keying in the following:

```
chmod +x radio_pulse.sh
```

In this script, you can also enable networking, but you might also find it very convenient to apply the paperfs program and select the option below:

'Make the discoverable PulseAudio network sound devices available locally'.

Lastly, you've got to set up authentication. PulseAudio makes use of a shared secret that's referred to as cookie for authentication. Simply copy the cookie from the Pi at **/run/pulse/.config/pulse/cookie to~/.pulses-cookie** in the home directory. It will now authenticate.

On your client, restart PulseAudio using:

```
pulseaudio -k && sudo alsa force-reload
```

Now run the above script; if everything goes on well, you will now hear the sound from your laptop's radio. You also have to make a couple of adjustments within the pavucontrol program. Make sure to turn the radio down squelch so that there is something that comes through. Next, click on the playback tab and then click show all streams and ensure the loopback from Pi is playing on gadget you want it to play on. This is what it should look like:

You can run pavucontrol for the Raspberry Pi as well from your client as follows:

PULSE_SERVER=raspberrypi pavucontrol

You can then adjust the Pi connections if you find it necessary.

You should ensure you set your device to DATA mode (or FM-D) if you are using an IC- 7100, or in a way that your USB audio gets into the transmitter. You can now do digital modes and voice through your network- how cool is that! Take a look at the image below that shows running fldigi on a laptop that is receiving audio in this setup:

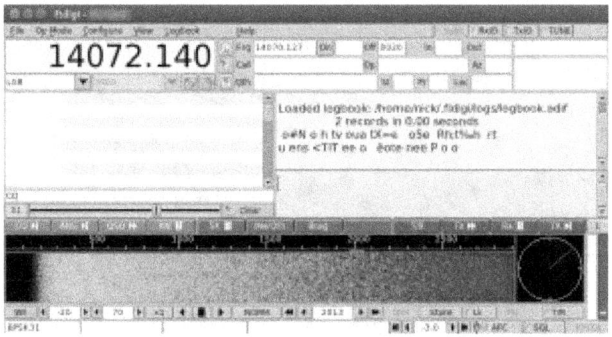

Remember that at this point, you are on a local network and so you'll have to create the VPN to go outside your house via the internet.

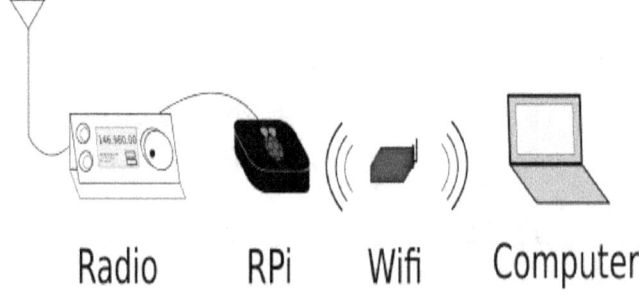

Control the radio from far

If you only want to conduct digital modes through fldigi, you probably are good to go at this point. You can run flrig on Pi and fldigi on the client, and they'll be able to connect and key up the radio and things like that. But what if you want to do voice as well, what do you do?

If you are a programming nerd, you can write your own small controller for IC-7100 in a language like Python. You'll find it simple, and will allow you to switch the radio on and off, switch the DATA mode on and off, access the different memory channels and key up the radio. This is a good start but you can make it a lot more sophisticated with time. You will particularly need to be able to key in the frequency you want or scan down or up the bands for HF; something that should not be too difficult to add. You can feel free to use it if you want but do not expect a lot; if you want to improve it, go ahead!

We have other programs out there offering the ability to control a radio via the network.

Troubleshooting

I personally went through quite a number of iterations and found numerous problems as I tried to figure it all out. I conducted the PulseRadio configurations on Pi at first but eventually realized that it was more flexible to perform it on a laptop. I tried so much to get downsampling to work to reduce the needed bandwidth. This is crucial for the slow Wi-Fi networks but particularly important when accessing your local network remotely via VPN through the internet. I did play a lot with resampling methods and reduced it to 11025 sample rate. It worked well and sounded fine for some time even though kept fading out after every two minutes. You can change your tunnel source to be the following (to downsample):

```
pactl load-module module-tunnel-source
server=tau.partofthething source_name=icom_source rate=11025
```

```
pactl load-module module-tunnel-sink server=tau.partofthething
sink_name=icom_sink rate=11025
```

This basically uses around 60KiB/s over the network and 200 KiB/s at default. You can change the sampling methods `edit/etc/pulse/daemon.conf` on Pi and adjust the `resample-method=speed-float-1` setting.

In some resampling conditions, it is possible for PulseAudio to use the Pi's CPU excessively. The main problem however, is that there are a few delay seconds as the audio moves across the internet.

You can then unload each of the PA modules onto your PC using the following command (this is applicable if you are playing around with different settings).

```
pacmd unload-module 32
```

You can use the pacmd list-modules to make a list of all the loaded modules where 32 is the module's index to upload. You can do this a bunch as you tweak your script and rerun it.

Connect to the LAN securely with a VPN

A VPN enables you to connect to your local network over the internet remotely. And you can configure your Pi to make this connection possible.

You can conveniently set up a 'road warrior' OpenVPN on your WRT-based router. This in turn will allow you to connect to you home LAN anytime securely from anywhere. You will find this useful especially if you want to print things from far away, avoid surveillance while on public or other untrusted networks or talk to your internal devices. This step usually entails securing your WRT-compatible router, then using ROM that has OpenVPN server such as OpenWRT or DDWRT to flash it then creating some private or public keys and then proceeding to configure the clients and server. This is not entirely difficult, but can take you quite some time to get it working as a result of the numerous instructions that are a bit different -which are present online. The GUI-based config in current DDWRT builds makes it very simple. To know more about building some RSA Keys and flash your router, please visit this guide and perhaps also this guide as well.

Congratulations! I hope this project works for you as it was very incredible for me to figure out.

Conclusion

We have come to the end of the book. Thank you for reading and congratulations for reading until the end.

The numerous facets of amateur radio have managed to attract practitioners with many different interests worldwide. Many amateurs start with a simple fascination of radio communications but with time find themselves combining it with other deep-seated interests of their own to make the pursuit of their hobby fulfilling.

Regardless of whether you want to pursue the focal amateur areas such as radio propagation study, technical experimentation, radio contesting, public service communication or computer networking, having an understanding, and skills on amateur radio are sure to change your life, and broaden your perspective on so much that you had never imagined.

The book could not have possibly covered everything you need to know in the world of amateur radio; but you have a footing now. Do more research and try to more explore projects you can do to become a better practitioner, with Pi.

If you found the book valuable, can you recommend it to others? One way to do that is to post a review on Amazon.

Click here to leave a review for this book on Amazon!

Thank you and good luck!